SPOTLIGHT ON PHYSICAL SCIENCE

ELECTRIC CURRENTS

IT'S ENERGETIC

BARBARA MARTINA LINDE

PowerKiDS press.

NEW YORK

Published in 2020 by The Rosen Publishing Group, Inc.
29 East 21st Street, New York, NY 10010

Editor: Elizabeth Krajnik
Book Design: Michael Flynn
Acknowledgments: Jeffrey A. Hinkley, Ph.D., Chemistry, subject matter expert

Photo Credits: Cover, p. 1 TWStock/Shutterstock.com; (series molecular background) pro500/Shutterstock.com; p. 4 Olivier Le Queinec/Shutterstock.com; p. 5 kryzhov/Shutterstock.com; p. 6 Photo smile/Shutterstock.com; p. 7 fen deneyim/Shutterstock.com; p. 8 Print Collector/Hulton Archive/Getty Images; p. 9 DEA / G. DAGLI ORTI/ De Agostini/Getty Images; p. 11 Universal Images Group/Getty Images; p. 12 Everett Historical/Shutterstock.com; p. 13 https://commons.wikimedia.org/wiki/File:Light_bulb_by_Lewis_Latimer,_1883_-_Museum_of_Science_and_Industry_(Chicago)_-_DSC06448.JPG; p. 15 Library of Congress/Corbis Historical/Getty Images; pp. 16–17 rumruay/ Shutterstock.com; p. 18 Christian Delbert/Shutterstock.com; p. 19 Sutiwat Jutiamornloes/Shutterstock.com; p. 21 Scharfsinn/Shutterstock.com; p. 23 https://commons.wikimedia.org/wiki/File:Apollo_SM_fuel_cell.jpg; p. 24 Taras Verkhovynets/Shutterstock.com; p. 25 Chones/Shutterstock.com; p. 27 PASIEKA/Science Photo Library/ Getty Images; p. 28 Gorodenkoff/Shutterstock.com; p. 29 Chris Minihane/Moment/Getty Images; p. 30 Peshkova/ Shutterstock.com.

Cataloging-in-Publication Data

Names: Linde, Barbara M.
Title: Electric currents: it's energetic / Barbara Martina Linde.
Description: New York : PowerKids Press, 2020. | Series: Spotlight on physical science | Includes glossary and index.
Identifiers: ISBN 9781725312999 (pbk.) | ISBN 9781725313026 (library bound) | ISBN 9781725313002 (6 pack)
Subjects: LCSH: Electric currents--Juvenile literature. | Electricity--Juvenile literature.
Classification: LCC QC601.L56 2020 | DDC 537.6--dc23

Manufactured in the United States of America

CPSIA Compliance Information: Batch #CWPK20. For further information contact Rosen Publishing, New York, New York at 1-800-237-9932.

CONTENTS

CHAPTER ONE

WHAT ARE ELECTRIC CURRENTS?

Imagine a world without television. Suppose you had to read by candlelight. Would you want to write all of your homework and letters on paper? What if there weren't any text messages or emails? With the flip of a switch, you have all these modern things. But you wouldn't have any of them without electric currents.

Electricity is a form of energy. The word "current" comes from the word *corant*, which means "running" or "flowing" in French. Currents are flows of electric charge. They may flow from a power plant or a battery. You can't see electric currents, but you can see how they work.

What happens when you flip on a light switch? What happens when you plug in a toaster? Why does a cell phone battery die? How do you charge it? Keep reading to find out how electric currents work!

Electric currents power many devices in your home, including smartphones and tablets.

CHAPTER TWO

WHAT IS A CIRCUIT?

For electric current to flow, it needs a complete circuit, or path. As shown below, the switch, the battery, the lamp, and the wire are all parts of the circuit. A circuit can be open or closed.

On the right side of the picture below, there's a loop from the battery to the light bulb and back to the battery. This circuit is closed. When the switch is on, electric current flows through the bulb and it lights up. The energy that was stored in the battery becomes heat and light. The bulb stays lit until the battery runs down. On the left side of the picture below, the switch is off, which is shown by the gap in the wire. Because the circuit is open, no current flows and the bulb doesn't light up. The energy remains stored in the battery.

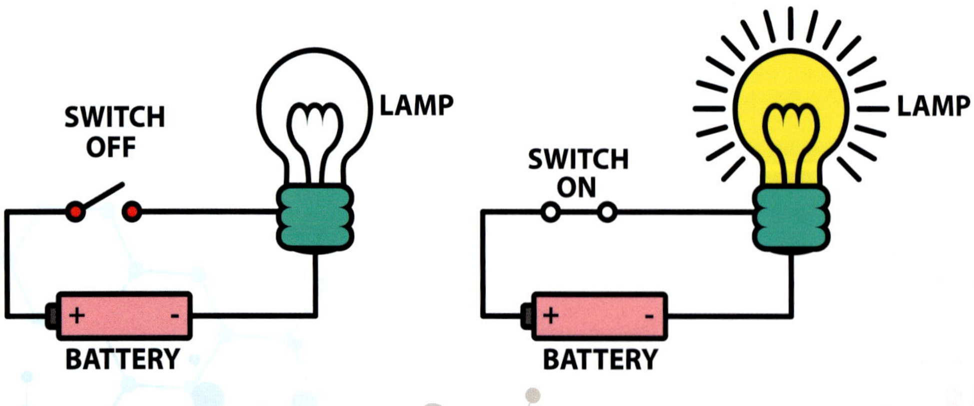

This picture shows an electric circuit. A wire connects the battery to a switch. If the switch is on, current flows through the switch and the light bulb. Then it flows back through another wire to the other end of the battery.

CHAPTER THREE

VOLTA AND AMPÈRE

ALESSANDRO VOLTA

In 1800, Alessandro Volta, an Italian **physicist**, invented the first electric battery. Known as the voltaic pile, Volta's battery was made of stacked alternating discs of silver and zinc. He put a piece of cloth moistened with an **electrolyte** between each pair. With this battery, scientists had a source of electric current that didn't need to be recharged. The volt, or unit for measuring the force that moves an electric current, is named after Volta.

Around 1820, André-Marie Ampère, a French physicist, researched the relationship between electricity and magnetism. Ampère found that parallel wires carrying electric currents attract each other if their currents flow in opposite directions or **repel** each other if their currents flow in the same direction. Ampère measured electric current. The ampere, or amp, the base unit of electric current, is named after him.

Ampère's discovery is quite useful. Every electric motor has coils of wire. When current passes through the wires, the magnetic force turns the motor.

CHAPTER FOUR

SAMUEL MORSE

Samuel Morse was an American painter and inventor. While returning from Europe in 1832, he learned about research on electricity and magnets. Back home, he experimented with batteries and coils of wire in a circuit. When he turned on the current with a switch, the coil was magnetized. He attached a pen to this magnet. By stopping and starting the current, Morse moved the pen up and down and made dots and dashes on the paper. These dots and dashes stood for letters and numbers in a system now called Morse code.

Later, Morse showed his invention in the U.S. Capitol building. He used miles of wire between the rooms. Congress gave him money to string wires between Washington, D.C., and Baltimore, Maryland. In May 1844, the line opened for public use. This was the beginning of modern communication. Within 15 years, telegraph wires stretched all across the country.

In Morse's original telegraph, pictured here, the switch interrupts the current's path from the battery to the coils.

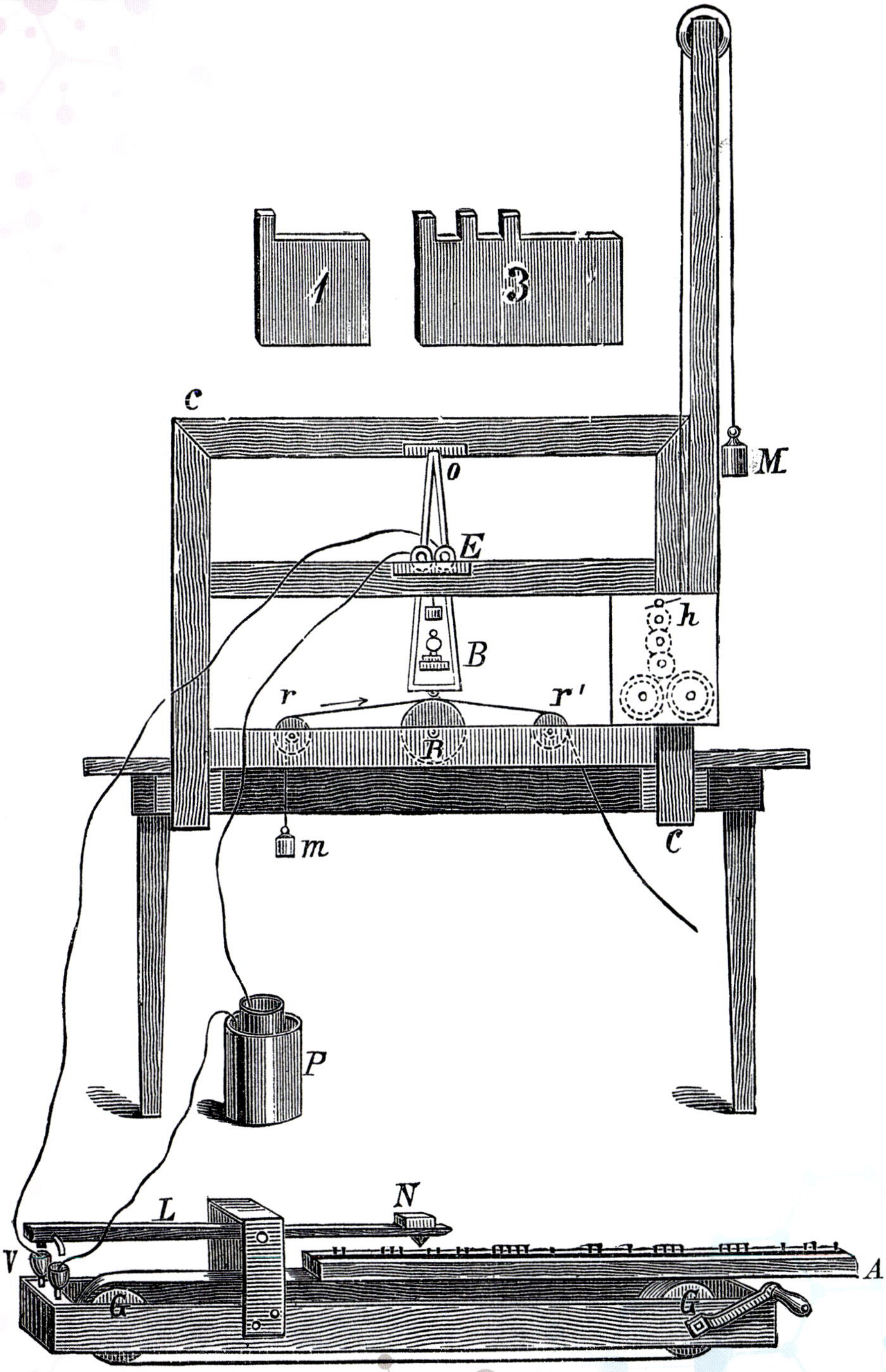
1
3
c
M
o
E
h
B
r
r'
R
m
C
P
N
L
V
A
G
G

CHAPTER FIVE

IMPROVING THE LIGHT BULB

In 1880, Thomas Edison filed a **patent** for an **incandescent** lamp, or light bulb, made of a carbon **filament** inside a glass globe. When an electric current passed through the filament, it heated up. The filament glowed red-hot, producing light. Edison's bulbs had very short life spans—sometimes just a few days.

THOMAS EDISON

In 1881, Latimer sold his patent for the improved carbon filament incandescent light bulb to the U.S. Electric Lighting Company. The light bulb pictured here is one of Latimer's bulbs.

Lewis Latimer was an African American inventor. He's known for creating a carbon filament that lasted longer than those in Edison's light bulbs. He created this new filament in the 1880s while working for the U.S. Electric Lighting Company. His improvement to the carbon filament made incandescent light bulbs last longer, and the materials he used made the bulbs cheaper. In 1884, Latimer left the U.S. Electric Lighting Company to work for Thomas Edison's power company. He helped set up electric plants in the United States, Canada, and England.

CHAPTER SIX

THE WAR OF THE CURRENTS

In 1890, Edison became one of the founders of the Edison General Electric Company, which became part of the General Electric Company in 1892. Edison developed direct current, or DC. This type of current only moves in one direction. George Westinghouse owned Westinghouse Electric. It used inventor Nikola Tesla's alternating current, or AC. This type of current changes direction from time to time.

The 1893 World's Fair was held in Chicago. Both Edison and Westinghouse saw the fair as a place to show off a type of electric current. Westinghouse won the bid to light the fair, but Edison and Westinghouse continued competing. Westinghouse beat out Edison to make the first **hydroelectric** power plant, creating power from Niagara Falls in New York.

However, AC could be sent over long distances, but DC could not. Before long, AC became the main type of power used.

The Electrical Building at the 1893 World's Fair was used to display electrical exhibits from a number of companies, including General Electric and Westinghouse.

CHAPTER SEVEN

THE POWER GRID

The electricity we use in our homes, schools, and businesses is produced at electric power plants. The plant may be hundreds of miles away. At the power plant, fuel is turned into energy. This energy is used to turn a turbine, or a large fan-like machine, which spins a **generator** to make electricity. Some power plants may use natural gas, nuclear energy, oil, or wind as fuel.

Electric Power Distribution

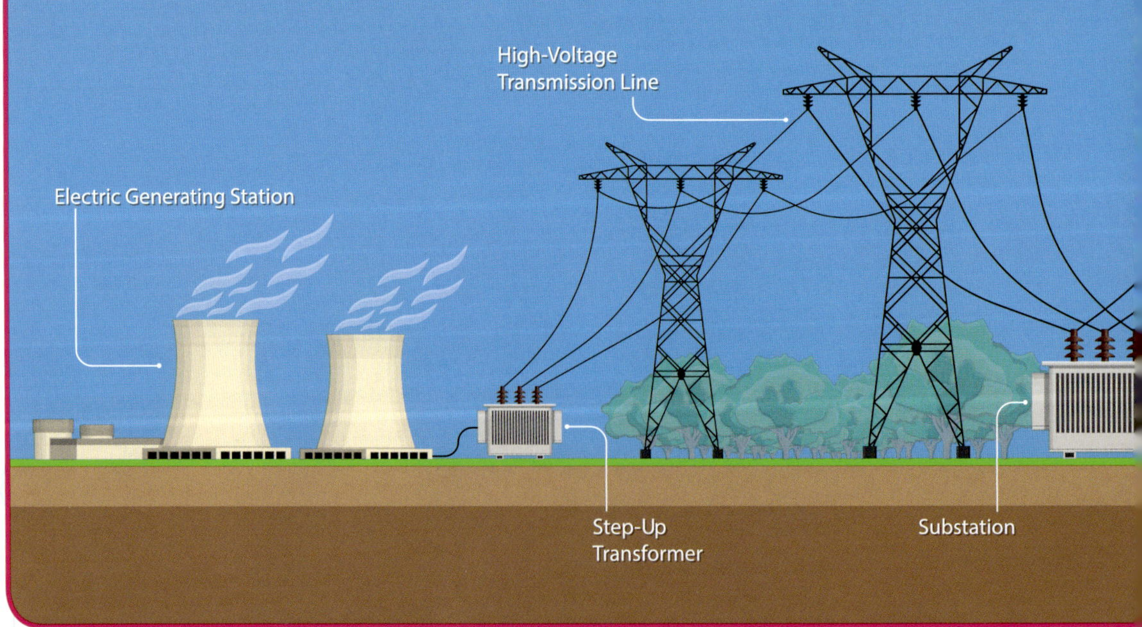

Wires on tall towers carry the electric current from the power plant to a substation. This may be close to your neighborhood. From there, the current is sent out to different neighborhoods. Wires may run down your street on poles. In some places, the wires are underground. This system of power plants and wires is called the power grid. Without the power grid, many of us wouldn't have electricity in our homes or businesses.

As of December 31, 2017, there were about 8,600 power plants in the United States. Groups of plants are connected to each other in subgrids. The grid system makes sure electricity can get to where it's needed.

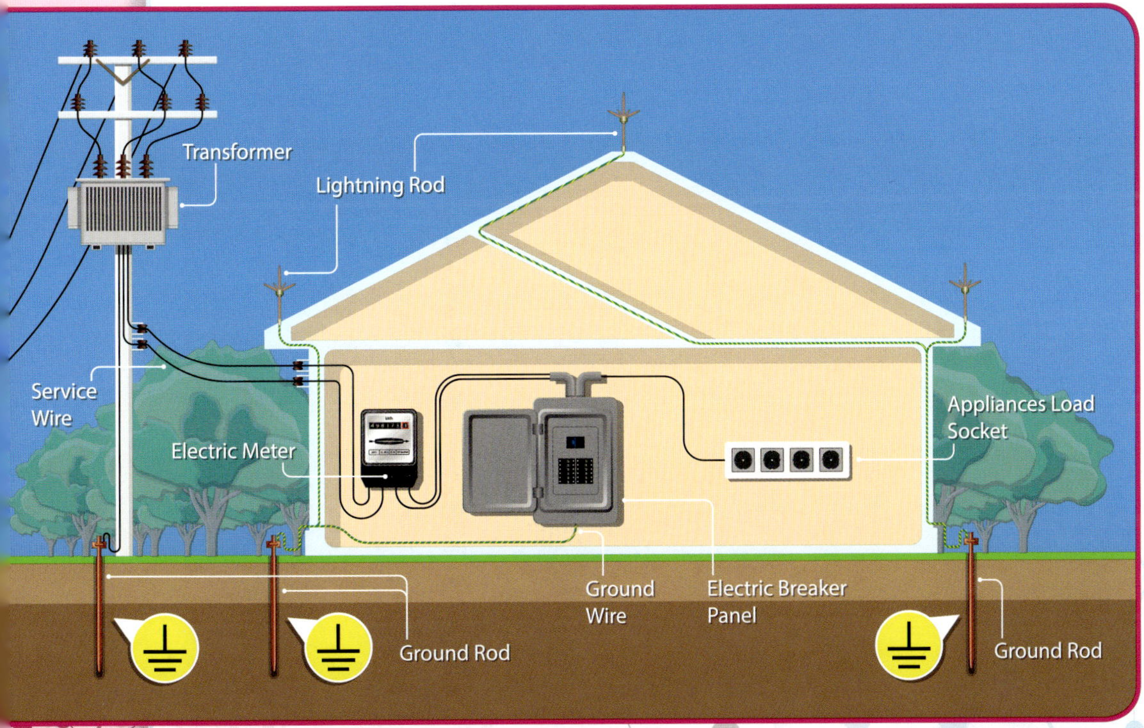

CHAPTER EIGHT

AROUND THE HOUSE

The wires that run to your house connect you to the power grid. The current goes through a meter on your house, which tells the electric company how much electricity you use.

Circuit breakers protect your home from overloaded circuits and fires.

Next, the current goes into a breaker box. A breaker is a switch that controls one electric circuit. When you flip a switch or plug something into an outlet, the current flows from the main supply. It goes through wiring in the walls to the light or outlet.

When you trip the breaker, or cause it to turn on, it opens the circuit to stop the power supply. Circuit breakers may trip because they're overloaded or because there's a short circuit. A short circuit is when an electrical current takes a short cut through part of the circuit. This can cause wires to get too hot and maybe even catch fire.

CHAPTER NINE

ELECTRIC MOTORS

Electric current powers things to light, heat, and cool homes. It can also power motors. Motors in homes and factories do work. At home, blenders, shavers, and even some toothbrushes use electric motors. In factories, motors inside machines do many different things. They shape metal, pump liquids, and move products. In 2018, the industrial sector used about 25 percent of the electricity produced in the United States.

Trains, streetcars, and subways use electric motors. They connect to the grid. Some use overhead wires. Others use an electrified third rail on the tracks.

Roads, however, don't have overhead wires or third rails. So, cars that have electric motors use batteries. Some cars run completely on electricity. Their batteries can be recharged at your home or at a charging station. The longest range of any electric car as of 2019 is around 300 miles (482.8 km). Hybrid, or combination, cars run on both gasoline and electricity.

Some cities have charging stations for electric cars to use when their batteries run low. There are charging stations along roads in some states and countries.

CHAPTER TEN

FUEL CELLS

In 1839, British physicist William Grove created the first fuel cell by changing the chemical energy of a fuel into electricity. He used hydrogen and oxygen as fuels that were energized by **electrodes**.

Fuel cells can produce a steady current as long as they get fuel. They never need to be recharged like batteries. Starting in the 1960s, NASA used fuel cells in space missions. The spacecraft didn't have to carry batteries, and astronauts didn't have to worry about batteries running out of energy.

New fuel cells can use air rather than pure oxygen as fuel. They can even use liquid fuel or **biomass**. Fuel cells still aren't a popular source of energy, though. They are often expensive to build and may not make energy as fast or as clean as other power plants.

A set of three fuel cells supplied power to spacecraft in the Apollo project that landed humans on the moon. The cells created electricity and water. The astronauts drank the water the cells created.

CHAPTER ELEVEN

BULBS AND LEDs

For many years, most homes and buildings used incandescent light bulbs. So did car headlights. Each bulb has a glowing filament. The current flowing through it heats it to almost 5,000°Fahrenheit (2,760°Celsius)! Even though these bulbs produce a lot of heat, they don't produce much light. **Fluorescent** bulbs are more **efficient**. Instead of using a filament, fluorescent bulbs pass electric current through gas, which causes the gas to glow. They produce the same amount of light as incandescent bulbs, but they use only one-quarter of the current.

LIGHT-EMITTING DIODES (LEDs)

From left to right, this photo shows an incandescent bulb, a compact fluorescent bulb, and an LED bulb. They are similar in size, but the LED shines the brightest and lasts the longest.

In 1962, an engineer at General Electric invented a light-emitting diode, or LED. This small electric device gives off light using a tiny amount of electric current. LEDs don't get hot and can be very small. Later researchers improved on them. White LEDs provide light for computer and cell phone screens. They use much less energy than incandescent bulbs.

CHAPTER TWELVE

TRANSISTORS AND COMPUTERS

Electric currents can provide light, heat, and motion. They can also move information. A transistor is a device that creates and controls electrical signals. A transistor can be a type of a switch and tell these signals where to go. It can also be a amplifier, which means it makes signals bigger. Computer parts use transistors to move bits, or small amounts of data. Electric currents take time to flow, though. Everyone wants their computer or phone to work faster. So, engineers are working to make devices faster.

One way to make devices faster is to make smaller microchips that fit more transistors. The closer the transistors are, the faster the device is. But you can only fit so many transistors on a microchip. Now, scientists are experimenting with quantum computing and artificial intelligence to solve the problem of slow computing.

This is a microscope picture of a computer chip. IBM plans to release 5 nanometer chips sometime in 2020.

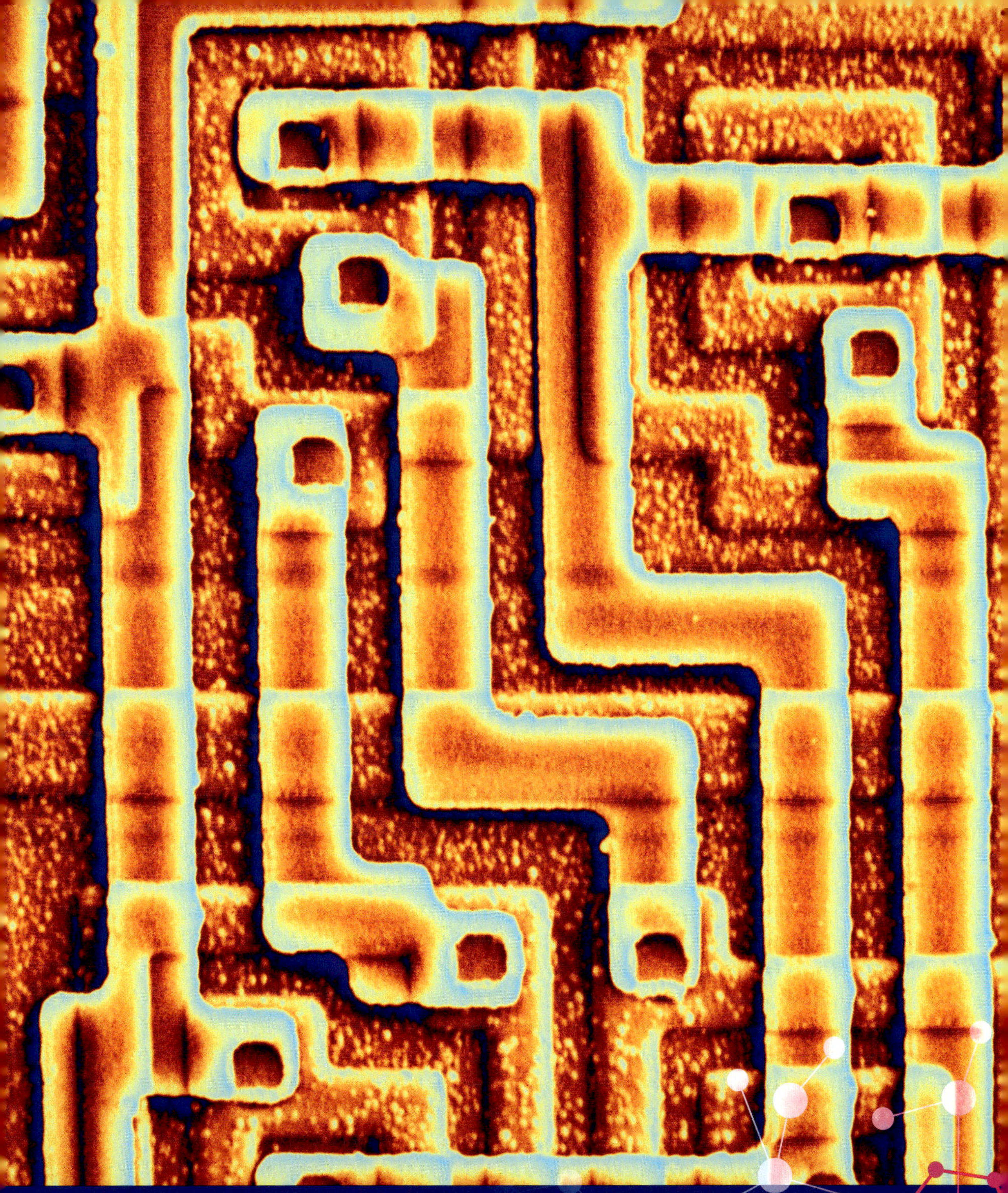

CHAPTER THIRTEEN

BE FLEXIBLE!

If wires that conduct electricity are thin enough, they're flexible, or capable of bending. They can even be woven or knitted into clothing. These could become part of a wearable computer. It could help athletes track their performance. It could send medical reports to a doctor.

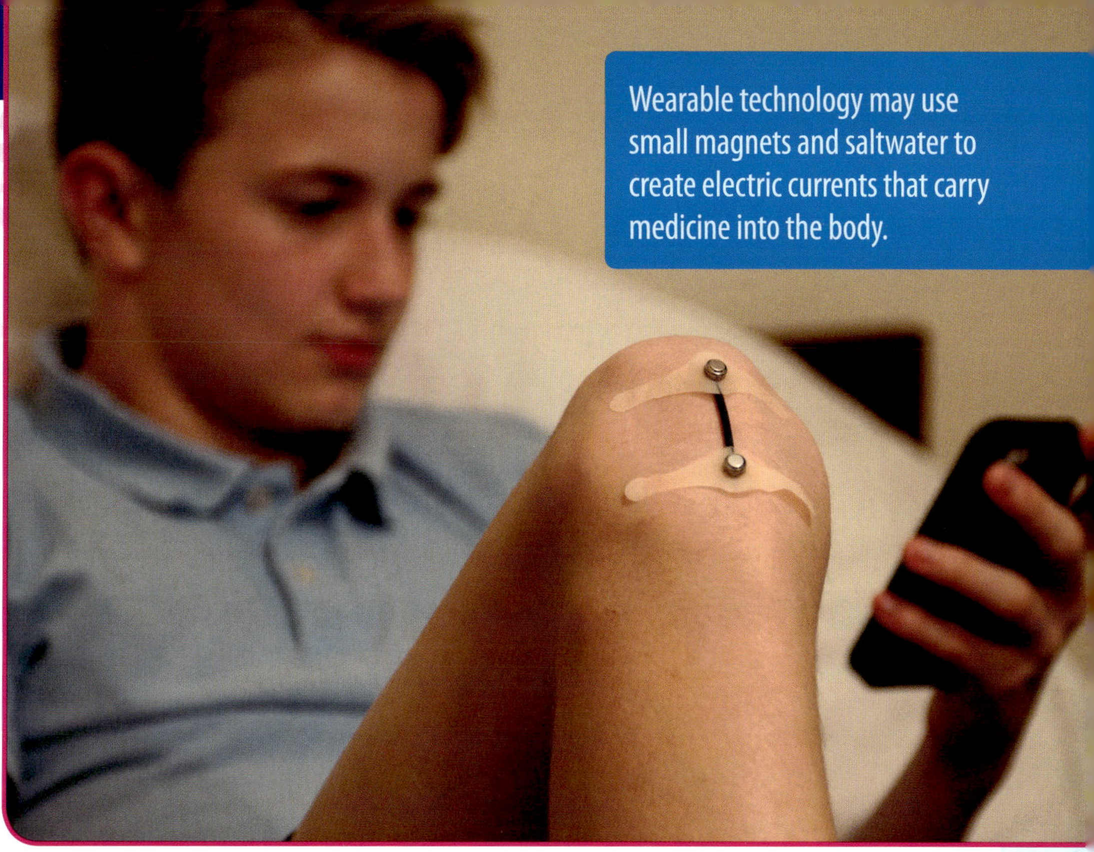

Wearable technology may use small magnets and saltwater to create electric currents that carry medicine into the body.

Another way to make flexible wires is to mix finely powdered metal with rubbery ink. The ink can be printed to make a circuit. After it dries, the ink becomes a soft wire. The circuit is stuck on the skin like a temporary tattoo. These circuits can report body temperature. Some can sense how much water is in the body. Others detect brain activity. Wearing the patch is more comfortable than long medical tests. A patch may also sense muscle movement. This could help people with physical disabilities move.

CHAPTER FOURTEEN

THEN AND NOW

If you were alive in the 1800s, you could send a telegram to a friend using a telegraph machine. Your house might have been one of the first to have electricity and use an Edison light bulb.

Most people living today have never lived without electricity. We turn on the lights and plug in devices to help us do work without even thinking about it. We have batteries for things such as flashlights. We depend on electricity for transportation, communication, and many other parts of modern life. It takes thousands of power plants to supply our need for electric current.

Look around you. Think about electric currents at work. Imagine the currents flowing from power plants to your city and to your home. Think about the scientists and engineers who brought electricity to us. Maybe you'll find new uses for electric currents!

GLOSSARY

biomass (BYE-oh-mass) Plant or animal matter used as a source of fuel.

efficient (ih-FIH-shuhnt) Capable of producing desired results without wasting materials, time, or energy.

electrode (ih-LEHK-trohd) One of the points through which electricity flows into or out of a device.

electrolyte (ih-LEHK-truh-lyt) A substance that when dissolved can conduct an electric current.

filament (FIH-luh-muhnt) A fine wire in a light bulb that glows because of an electric current.

fluorescent (fluh-REH-suhnt) Producing light when electricity flows through a tube filled with a type of gas.

generator (JEH-nuh-ray-tuhr) A machine that uses movement to produce electricity.

hydroelectric (hy-droh-ih-LEHK-trik) Relating to or used in the production of electricity by waterpower.

incandescent (in-kuhn-DEH-suhnt) Producing bright light when heated.

patent (PAA-tuhnt) Rights given to a person by the government to make, use, or sell an invention.

physicist (FIH-zuh-sihst) Someone who studies physics, or the science that deals with matter and energy and their interactions.

repel (rih-PEHL) To force something to move away or apart.

INDEX

PRIMARY SOURCE LIST

Page 8
Alessandro Giuseppe Antonio Anastasio Volta. Painting. ca. 1827. Now held by Hulton Archive on Getty Images.

Page 9
André-Marie Ampère. Painting. Now held by DeAgostini on Getty Images.

Page 12
Thomas A. Edison in his West Orange, New Jersey, laboratory. Photograph. ca. 1901. Now held by Everett Historical on Shutterstock.com.

WEBSITES

Due to the changing nature of Internet links, PowerKids Press has developed an online list of websites related to the subject of this book. This site is updated regularly. Please use this link to access the list: www.powerkidslinks.com/SOPS/electric